5⁰⁰

D0923764

Dying Forests – a crisis in consciousness

Dialogue with nature

Transforming our way of life

Pictures and text by
Jochen Bockemühl

Translated by John Meeks

Introduction by Brian Goodwin

Introduction

One summer's day about two years ago, a friend and I took a rambling walk through the countryside in the neighbourhood of the village where I live, Aspley Guise, in Bedfordshire. We were discussing perennial themes, but they centred on the difficulties encountered by biologists in understanding the particular type of process that is at work in living organisms. The source of these difficulties seemed to be the persistent tendency in modern biology to look for a special controlling centre in organisms, a clockwork within the orange, which is the cause and essence of all their properties of growth and regeneration, form and transformation, responsiveness and behaviour. This essential cause is currently located in the genome, which is considered to contain all the information required to generate the living, functioning organism. That is to say, the organisms we see about us are the artefacts of a special substance within them, their DNA.

This dualistic description of process, splitting it into cause and effect, is an old trick in Western thinking, and is just like the mind/body dualism of Descartes that has caused so much difficulty in understanding the relation between thought and action. It is the divide and conquer strategy of the analytical tradition, extremely effective in taking things to pieces so that the various aspects of their activity can be identified and manipulated, but very deficient at giving an understanding of whole, integrated process. The worm at the heart of the reductionist Cartesian philosophy of substance is the proposition that complex wholes are made up of simpler, self-sufficient, inert parts whose properties can be studied independently of, and can explain, the whole. The result is that all of nature falls apart into dead mechanisms.

Our ramblings took us across open farm-land, neatly hedge-rowed, and into the magnificent mature woodland of Woburn Abbey. Beech, oak and chestnut mingled with ancient pines, while rabbits and munt-jack deer slipped quietly away into the cover of bracken and elder. Pheasants leapt out of the undergrowth and went whirling off with great squawking. We considered how to conduct an enquiry into living process that preserves the insights of analysis, but holds the constituents together as a necessary unity. There is no denying that reductionism works, up to a point, especially in the study of relatively simple, isolated systems. However, it fails badly when applied to the dynamics of complex wholes, and we can see the destructive consequences all around us. Intensive agriculture over-extends itself in the direction of monoculture and dependence on special crop varieties, artificial fertilizers, herbicides and pesticides, pushing dynamically balanced ecosystems into patterns of run-away ecological damage. Industrial wastes add to the imbalance, and we now have the large-scale phenomenon of dying forests. My friend and I were acutely aware that the forest through which we wandered was protected from this destruction, not by good management but by good luck: little industrial pollution affects this part of England.

This is not so of many other parts of Europe, and of North America. In his beautiful and moving book "Dying Forests", Jochen Bockemühl urges us to see this phenomenon as evidence of a crisis in consciousness, in our way of looking at natural process. The book is richly illustrated with his own paintings of European landscapes, forest glades with their profusion of flowers, farm-land and meadows. They show the changing mood and texture of complex, living patterns with season, with good and bad husbandry, and under the action of destructive environmental influences. The book is permeated by a remarkable capacity to read the signs, to reconstruct the history of a tree or a copse or a landscape from its present state, to understand what intrinsic and extrinsic influences are expressed in the form of a plant or a community. This comes from a life-time of practical work, and the continuous development of an alternative scientific system to that of reductionist Cartesian analysis.

The alternative tradition within which Bockemühl's work is located is itself perennial, but the specific historical influence that shapes the research comes from Goethe's conception of natural process and his studies of plant form and transformation. For Goethe, nature was essentially and intrinsically dynamic so that movement and change are of the essence. He showed that the various parts of plants, such as leaves, sepals, petals and anthers, can all be understood as transformations of one another. During the growth of the plant, each of these parts arises from a bud, and initially all buds have much the same conical shape. However, depending on their position in the plant and its age, these buds develop in different ways, as Bockemühl and others have shown. Thus the growth of the whole plant is a unified process in time and space that realises itself in a series of parts with characteristic form, each related in a systematic way to the others. The particular forms expressed in these parts, and hence in the whole plant, are influenced by both internal factors (such as genes) and by the external environment. However, there is no legitimate way in which the form of the plant can be reduced to a set of primary causes from the genes. This would be to ignore the dynamic of the developmental process, which involves an irreducible space-time order. All three components, genes, development and environment, are in fact inextricably united in generating a plant, and there is no preferred causal level. There is plenty of evidence from contemporary biological research that this 'Goethean' view is in fact a much more satisfactory way of

understanding biological process than the alternative, so that the work of Bockemühl and his colleagues is in no sense peripheral to mainstream biology. There is now a movement within the subject to develop essentially the same conceptual orientation and associated research programme, using modern techniques and mathematical analysis.

Bockemühl's approach to the broad spectrum of urgent problems with which we are now faced is a part of Rudolf Steiner's anthroposophy, within which Goethe's original insights are interpreted and elaborated. This provides a comprehensive framework embracing all aspects of being, not just the biological and the physical. Thus, Bockemühl 'reads' nature not in the circumscribed manner of the naturalist, who is also sensitive to history and present process, but in relation to a total ontology of spirit, soul and substance. In his world, everything is connected by an inner necessity, resulting in patterns and gestures in nature that are simultaneously natural forms and symbolic expressions. Thus the phenomenon of the Dying Forests is an eloquent statement of dis-ease and disharmony throughout the realm of being, in thought and imagination as well as in action.

The recovery of ease and harmony proposed by Bockemühl is eminently practical. In the context of husbandry of resources, this takes the form of what is called biodynamic agriculture, and the development of appropriate technologies of the type that Schumacher did so much to encourage in his book "Small is Beautiful". At the conceptual level, the revolution that is now gathering momentum is dramatic in its implications, whether or not we accept the anthroposophical context within which Bockemühl's work is located. It involves recognising that parts and wholes have equal status, so that there are no preferred levels of causal explanation;

nothing is trivialized because everything has its value; subjects and objects are interchangeable, so there are no privileged observers. This gives us a true ontology of process and transformation. The world in its totality then becomes alive again, and we within it. Goethe's own description of this dynamic world of process has a renewed validity: "What has been formed is immediately transformed again, and if we would succeed, to some degree, to a living view of Nature, we must attempt to remain as active and as plastic as the examples she sets for us". This, concretely illustrated and sensitively developed, is also Bockemühl's message.

Professor B. C. Goodwin
Biology Department
The Open University

© Copyright 1986.
English translation by Hawthorn Press,
The Mount, Main Road, Whiteshill, Stroud, Gloucestershire GL6 6JA, U.K.
Translated from the German Edition © 1984 published by Philosophisch-Anthroposophischer Verlag am Goetheanum, Dornach, Switzerland, as Sterbende Wälder – eine Bewußtseinsfrage.
ISBN 1 869 89002 7
Typeset by Glevum Graphics, 2 Honyatt Road, Gloucester. GL1 3EB
Printed by Bath Press

All rights reserved. This book is sold subject to the condition that it shall not, by way of trade or otherwise, be lent, re-sold, hired out, or otherwise circulated without the publisher's prior consent in any form or binding or cover other than that in which it is published and without a similar condition including this condition being imposed on the subsequent purchase.

Contents

This book originally came about as a contribution to a "Special Exhibition on the Forest" under the auspices of NATURA 1984, in Basle. In this exhibition various bodies such as Swiss government organisations, university institutes, forestry and conservation groups, showed vividly the phenomena of the dying forests.

For this reason, when asked to make a contribution from the Goetheanum, we were able to assume a basic knowledge of the facts: the outer symptoms of illness, the pollutants in the soil and atmosphere which are held responsible, and the obvious counter-measures which suggest themselves. These things are summarized today in numerous books and articles.

I was not so concerned with the most obvious forms of criticism. It was rather my intention to show how factors connected with the whole life situation of humanity underlie the outer dying in nature. This is where the transformation must begin. Only where a conscious encounter with nature takes place can the values of life be seen from a new perspective. Then the strength can grow to give up deeply ingrained habits, and to bring about real changes in the area of technology. This requires courage. Once this is understood, the exercises offered here will be a contribution to nature as well as humanity. On the quest for panaceas, and in our eagerness to do something here and now, this side is too often forgotten, although it affords the only possibility of giving any real endurance to measures which we must naturally take without delay.

I would like to express my gratitude to all those who have commented on this book, both positively and negatively. It is our continuing concern to find new approaches to overcoming these problems on different levels. So we can hope in the course of time to make further contributions in this field. Nonetheless, it must remain the concern of each of us in the end to find what individual contribution we can make out of our own life situation.

In our courses on landscape ecology in the study programme of natural sciences at the Goetheanum, we observed selected natural habitats both in Switzerland and abroad. On the following day, we always attempted to reconstruct a shared picture of the place. We did this with the help of individual plants which we had brought back with us, or which we remembered. Blackboard sketches were made. As a next step, we tried to commit to paper the impressions we had formed. The essential task was always to capture something of the overall mood of the place, so that the ideal unity could speak through selected details. Sketches made at the site were intended only to reinforce our observations. They were seldom used directly for the final painting.

Studies of this kind are suited to encourage the development of an enhanced sensitivity to the creative as well as the destructive forces at work in the natural and human world. The dying forests are but an especially striking symptom.

Against this pictorial background – which was the fruit of individual experience – it is far easier to approach the study of details, or to conceive possible directions for experimental work, than without this support.

I am grateful to all who have collaborated in this effort, or otherwise helped to make it possible. A particular word of thanks is due to Dr. Georg Maier and the other co-workers at the Goetheanum Research Laboratory.

The studies by Max Moor (Basle) have been of great value in planning the excursions in this area.

Jochen Bockemühl

Dying forests – a crisis in consciousness

The forests are dying.
Who is responsible?

* Sulphur Dioxide, Ozone, Nitric Oxides?
* Industry, Power-plants, Motor traffic?

Only human beings can bear responsibility.
But who?

* The scientists or engineers who developed the motor cars and power-plants?
* The operators of machines designed to satisfy human needs?
* Politicians?
* We ourselves as consumers and beneficiaries?

We all have our place somewhere within this network of shared responsibility. We have the feeling that we are but links in a chain of circumstances which we are powerless to influence. Where does responsibility arise? Where are we free to take an initiative of our own?

Figure 1:
A future vision for our forests, which has already become a tragic reality in the Erz Gebirge in Bohemia (Czechoslovakia and the German Democratic Republic).

Life with the forest – a retrospect

In earlier times man lived with all his thoughts and habits of life in intimate communion with the course of the day and the year. He experienced nature more from within; consequently, his practical activities were guided by healthy instincts. Natural reverence towards the forest guaranteed its preservation.

Figure 2:
Even today one can still find landscapes which have been shaped traditionally in a richly differentiated way. So for instance in the Swiss mountains, or here, in the Lower Bavarian Forest.

The decline of cultures has repeatedly followed in the wake of forest destruction.

Such waves of destruction are always accompanied by a loss of a view for the whole in timber consumption, land clearing and pasturing. Such is frequently the case today in developing countries, where old traditions are falling into decadence. So, for instance, in the Amazon rain forests of Brazil.

Figure 3:
Landscape in Delos (Greece).
Temple ruins in a treeless landscape which was once covered in forest.

With the emergence of the new age of science, a new and intensive wave of destruction was unleashed against nature.

Exploitation no longer took place to fill domestic needs, but in order to develop railroads and industries. Gradually, as the consequences became apparent, concerns were expressed about the loss of the natural environment. Trees were planted. Economic considerations, however, channelled this activity into the development of fast-growing monocultures for timber production. 'Economics' ignored the cultivation of living, differentiated forests.

Figure 4:
Monotonous spruce plantations in Northern Germany. Through such practices, man unwittingly brought on natural calamities, such as epidemics of woodland pests.

Independently of these events, ideas on the shaping of landscape were developed at the beginning of the last century.

A new awareness arose for the loss of natural habitats. The idea of nature conservation was born, and ecological considerations were brought to bear on landscaping.

In principle, these ideas made possible effective work to counterbalance the direct destruction of nature when architects, landscape architects, farmers and foresters collaborate in the right way.

Figure 5:
The boundary between natural, well-thinned beech forest and a spruce forest unsuited to the locality. The poor soil of the spruce plantation allows virtually no ground vegetation to become established.

The latest threat has deeper roots.

Our modern forms of life and thought have given rise to consequences whose destructive influence on the forest and on ourselves has assumed the dimensions of an elemental power. Our habitual thought-patterns with their linear chains of outer cause and effect are no longer adequate to meet this threat. It is not enough to look for the outer symptoms in exploitation or false economic thinking. We are dealing with a problem which has its roots in our own conscious relationship to nature.

To be sure, constructive first steps can be taken before these deeper questions are tackled. After a critical appraisal of the situation, one can move on to seek alternatives in dealing with specific technical problems. One form of technology can replace another, and so alleviate the worst destructive influences. But our energy requirements are not yet reduced in this way, and without a fundamental change in our whole way of thinking, the side-effects still cannot be foreseen.

Dying forests and air pollution, seen in this context, are but symptoms of the loss of our instinctive connection to nature on all the various levels of experience, understanding and action. How can we find our way back to the interrelated whole, as a prerequisite to assuming responsibility?

Figure 6:
Nuclear Power Plant

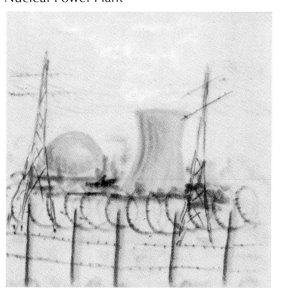

Figure 7:
Solar Energy Collector

What can we do?

The avenues of escape which people seek in the face of the environmental crisis may not represent realistic solutions; but they can nevertheless point a way to overcome the fundamental split in our consciousness. The first step must undoubtedly be to work outwardly against environmental pollution:

* On a large scale, through political action and laws, through reforms in technology, reduction of exhaust fumes, etc.
* On a small scale, through manifold efforts to economize and to conserve.

But by itself this will not be enough. Economic and social obstacles stand in the way of legal reform, and economizing cannot become the sole object of life. In the face of such difficulties, we find ourselves hemmed in by constraints which, when projected into the future, give birth to fears. So overwhelming is the sense of helplessness, that an unconscious longing for escape is the inevitable result.

But we can only 'escape' the consequences of our attitude to life by transforming this attitude.

Figure 8:
View from Mount Rachel in the Upper Bavarian Forest.
Here, in a national park, far removed from industry and civilisation, many people seek refuge in the 'wholeness' of nature. But the dying of trees proceeds at a threatening pace.

Modern man feels himself to be an external observer. He looks out at the world, and his questions probe no more deeply than to outer chains of cause and effect. He discovers natural laws of universal applicability. He learns to manipulate them freely, making them the foundation for technology.

Whether or not he allows ethical values to enter this process is his own personal affair. Here lies the danger of our modern situation. Inner moral experience has been sundered from the understanding of natural law.

Our newly-won freedom, together with the development of unsuspected technical possibilities to satisfy conscious and unconscious desires, has led to the domination of technology in our surroundings. Forces have been unleashed which have taken on an existence of their own, isolated from the natural context of earth and cosmos alike.

To an increasing degree, the unforeseeable side-effects of modern technology on nature and society have given rise to problems. In earlier times we found our security in the health of nature. Today security is sought in the so-called 'health' of industry.

What is happening in nature is more than ever dependent on our attitude to life – but inwardly we have cut ourselves off from nature.

The urge to escape

The radical split between our own inner life and the outer world is a universal experience today. Each human individual can discover in himself a region of inner experience which has nothing to do with the outer world. Correspondingly, in the outer world he can only find what he himself is not. Man cannot endure such disparity. He longs for escape. Two possible avenues seem open:

*   One direction of escape lies in the denial of the spirit, which is allowed validity only as an emanation of the external, material world. But by denying the spirit, man denies his own true being.

Such an attitude to life, allied to Darwinism, underlies numerous attempts to derive the "Origin of the Spirit" (Hellmuth Benesch) – that is, the origin of consciousness and the psychic functions – from material processes. Matter, be it understood, is in this view no better than 'dust' – a concept devoid of all content and quality. The adherents of such theories overlook the fact that their own standpoint can only be gained with the help of thinking. The activity of thinking in its own right can only be grasped inwardly. To take hold of this experience intuitively can give the certainty that thinking is an activity of the spirit. But there is in scientific circles a general reluctance to enter into the realm of introspective experience, even though the results of thinking are applied without reservation. So long as this remains the case, true morality will be unattainable. What remains is the 'So-Called Evil' as described by Konrad Lorenz.

This unbalanced path of scientific endeavour thus leads ultimately to a loss of identity, and to irresponsibility.

* The alternative avenue of escape lies in the quest for meaningful experience on a level which has nothing in common with the outer world.

Personal experience takes precedence over all that could be learned through the study of natural processes – and this leads in the end to a virtual enmity to science. The quest for knowledge is renounced in favour of retreat into an inner world. Drug-addiction, yearning for the East, or an escape into literary fantasy are fruits of this longing.

As guidelines for practical life, one is left with the concepts which have been implanted by our modern educational system. To be sure, they are products of scientific research, but for most people they have become so abstract that their origin is no longer understood. Who, for instance, when he speaks of 'Phosphorus' is mindful that he is speaking of the 'light-bearer' among the elements? Or of what he means in speaking of 'substances' and the 'radiations' which they emit? There is little openness to the insights which can be gained when the quality of thinking, appropriate to a given situation, is fructified by observation. And so the belief arises that such isolated concepts already represent a full reality.

The superstition and poverty of real content to which this leads are equally the consequence of an unbalanced development of scientific thinking.

Unquestionably, the unconscious urge to escape represents a danger, but at the same time an opportunity to re-establish our true connections with the world.

* Inherent in materialism is the striving towards selfless devotion to a reality outside ourselves.

* When we withdraw from the outer world, we are motivated by the quest for an essential aspect of reality within ourselves which we seek to attain through inner experience.

For the individual, inner disparity and conflict are the usual products of the unconscious mingling of these two urges to escape. It is particularly striking in the visual arts and the theatre, where the endeavour is made to expose it in the most manifold ways.

But there is a possibility to reconcile these tendencies, when we allow genuine experience to arise selflessly through sense-perception, and so develop a sense for what the phenomena themselves would tell us. This is the foundation of Goethe's scientific method.

Goethe has shown how a natural phenomenon may be investigated in such a variety of ways, that something of its essential nature can become directly accessible to outer observation. He has demonstrated this particularly clearly in his colour and plant studies.

Such careful and patient observations of the outer forms of a living, growing plant can thus penetrate to an experience of its inner nature. We all make use of this faculty. But it can be raised to consciousness and developed.

To clarify what is meant, let us look at the leaf forms of a plant.

Leaf forms of a field poppy (arbitrarily arranged)

These leaves stand in an outer, spatial order which in this case we have established arbitrarily. But we can also become aware of an inner order by attending to the 'similarity' – that is, what we can discover through our sense of relationships – between the ideal identity (field poppy leaves) and the outer differences.

Whether or not one already knows something about plants, or is familiar with the field poppy, makes little difference; after carefully sorting the leaves, everyone will sooner or later arrive at the same arrangement.

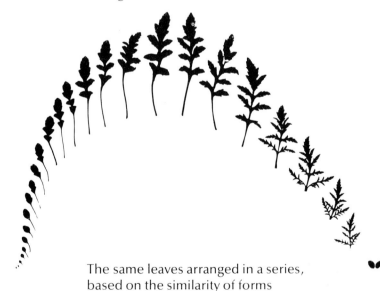

The same leaves arranged in a series, based on the similarity of forms

Through observation which is at once exact and artistic, we can thus discover an underlying law, out of which the plant – quite independently of us – produced the leaf forms, one after the other.

In the connection between the outer percept and the inwardly-produced concept, we step fully into reality on the levels of knowing and perceiving at once.

The natural sequence of the field poppy leaves from below upwards in the plant.

Figure 9
Organs of a hybrid Peony
The fully developed leaves, petals, stamens and ovaries are so arranged that the metamorphosis described by Goethe can easily be followed from one level to the next. The qualities of form and colour which we can discern here within the overall movement of expansion and contraction, are not the result of an outer process. They are the expression of the formative laws of the Peony, and can only be grasped by an imaginative, pictorial faculty. They belong to the natural, living context of a pictorial sequence which embraces the individual percepts.

At the beginning of this century, Rudolf Steiner showed concrete possibilities to bridge the gulf between worlds of inner and outer experience. He described Anthroposophy, the fruit of his life's work, as a 'path which would lead the spiritual in man to the spiritual in the universe'. Many practical spheres of life have been fructified by Steiner's teachings: among them education (the Waldorf School Movement), curative education, farming and gardening (biodynamic agriculture), medicine, social development and the arts.

The world centre for the anthroposophical movement is at the Goetheanum, Dornach, Switzerland. Research is aimed at a deepening of our understanding of the world. This involves a schooling in thinking and observation. For the process of knowledge is always related to the whole, and is not looked upon merely as an end result which can be studied and taught by itself.

The most essential task of the scientist must surely be to seek for what the human soul needs most in its quest for knowledge: an intimate relationship to the most manifold qualities of nature. The discovery of such qualities cannot but lead to the discovery of what is spiritual in the world around us. But to find the spiritual in nature, it is not sufficient either to retreat into a private world of subjective thought and experiences, or to lose oneself in objective dedication to the world of percepts. In either case the link between the inner and outer world is not maintained with sufficient wakefulness, and therefore makes us unfree.

The quest to understand the true nature of the world around us leads to the quest for our own true nature. Only from this vantage point can we hope to build a meaningful bridge between the worlds of inner and outer experience, and so maintain ourselves in face of the gap which separates them.

In what does the unity of a landscape consist, which we vaguely apprehend, yet can hardly define?

When entering a foreign country, we are apt to become more conscious of the landscape than when surrounded by familiar scenes.

We try to take in the scene with all its details, and then to connect them with our first impression. In this way we discover interrelationships. What we took in vaguely with our first impression acquires substance and content.

Figure 10
Morning mist over a landscape in Sussex (Southern England). In the early morning we have sought out a favourite spot. Mist still hovers over the land. The meadow and a few trees are dimly visible; in the foreground stands a large tree with rigid branches. But its contours are softened by the delicate red and green tones of budding foliage. Somewhere nearby a blackbird is singing.

It is normally through chance that we receive impressions of this kind. Outer facts and relationships mingle with aesthetic experience. What is their meaning for the reality?

In three fundamental ways we can begin to train ourselves to understand and experience natural relationships.

* First, through the development of a sense for facts.

We are already well-practised in recognising how we can make use of the things around us. Utilitarian thinking enhances our self-assurance. But only by turning our attention outwards, away from ourselves, can we begin to apprehend things in their own right.

* Our sense of beauty leads us to a sense for interrelationships within the whole. But if we surrender too completely to our feelings, we are in danger of lapsing into a wishful dream-world. Our sense for facts can help us maintain a balance. In alliance with the aesthetic sense, it can help us recognise the latent possibilities for development.

* In seeking the balance between the factual and the beautiful, we can begin to develop a sense for the interests of the entities around us. In fashioning nature to our service, we learn to consider not only our own needs, but those of our fellow beings in all the kingdoms of nature.

Landscapes: three approaches to an understanding of the whole.

1. The present context of phenomena.

This little wood has grown naturally over a heap of stones in an orchard. In such a 'miniature forest', we can discover three ways in which the single phenomena are interrelated:

* By the illumination and distribution of light and shade.

* By the tendency of the trees and shrubs to blend together into a unity. Oaks, hornbeams, ashes, field maples, sloe, hawthorns, whitebeams and many other plants and shrubs – all strive towards this overall unity of their own accord. The farmer only keeps them in bounds through the clearing of the meadow.

* To the north (right) the growth is generally long and slender, to the south it is thick and stocky. Here the variety of species is also greater.

We thus become acquainted with the natural context which is revealed to us by the weaving, spatial activity of light.

Figure 11
A little wood growing over a heap of stones collected from the field on the edge of the Table-jura landscape near Dornach, Switzerland. The context of phenomena becomes visible in three nuances.

Earthly Forces

When we encounter the objects of the outer world, we experience through them that part of reality which we ourselves are not. This is the foundation of our objective consciousness. In thinking about the forces of attraction and repulsion, through which these outer objects affect one another, we bring them into relationship with bodily experiences acquired through our limbs. These are the laws of the earthly forces.

Cosmic forces

In the manifold pictures of the phenomenal world, we experience relationships within a totality. They can lead us to what is ideal, beautiful. This is the foundation of our consciousness of the cosmic forces. 'Cosmos' means literally: the beauty, order, ornament of the universe.

Beauty is but the expression of a principle universally active in nature. Each little group of trees or plants, left to itself, strives out of its own inherent forces to assume a rounded, but rhythmical form, and so to blend harmoniously into a larger life-context. In untouched nature this can lead to a certain monotony. By consciously directing certain themes or motives from nature, man can raise nature to a higher level. In a forest, for instance, the Oak can only develop its characteristic form to a limited degree. The typical oak comes to be when man restricts the vegetation around it. Such practices are guided by a deeper interest for our natural surroundings. Their consequences for the inner life have yet to be fully appreciated. They bring about a healthy increase in the productivity of nature; and in many respects they provide a more important source of nourishment than what we take in physically.

Bio-dynamic agriculture

In earlier times, beautiful landscapes were a natural result of man's work with the land, so long as he was guided by healthy feelings and instincts. Today one can appreciate the laws out of which they were shaped, and try to protect and conserve them. But can the scene be preserved when its vitality is diminishing?

There is, however, another possibility. By developing a sensitivity to the forces which affect the totality of a place – that is, to the cosmic forces – one can allow oneself to be guided by the question: What would like to happen here? One is no longer working with a finished plan from the outset. One is striving to participate in a dynamic, living process.

The idea of the place for which one has assumed responsibility becomes the goal, the totality which seeks, through the hand of man, to become a reality.

In this sense, Rudolf Steiner has shown how agricultural practices can be brought into conscious harmony with nature, so that man's justified interests can help to shape the totality of a farm in an individual way.

2. The rhythmical relationships of natural processes (context of transformation).

Each landscape scene we observe leads us of its own accord beyond the bounds of the present moment. Features of its history begin to speak through the phenomena.

These wintry trees in a poplar wood in the former flood plain of the Rhine have undergone a development. The leaves which allowed this to happen are no longer visible. Growth-segments in the twigs and branches reveal the rhythmical traces of past seasons. The buds, which are already beginning to swell, tell of the future flowers and leaves. The dry reeds in the background also point to past vegetation and help us to imagine the new growth to come.

Figure 12
Wintry Black Poplar and Alder Forest in the former flood-valley of the Rhine near St Louis (Basle). One sees remnants of the past and anticipates something of the future.

These views from the Gempen Mountain near Dornach in the Swiss Jura, are painted at different times of the year, to enhance our consciousness of the seasonal qualities. At any given time, we can see only one particular scene.

Figure 13
View from the Gempen Mountain: Spring.

The other scenes must be recalled from memory. The essential thing is that we learn to experience the transformation in the course of time. We could compare this to a musical experience.

Figure 14:
View from the Gempen Mountain: Summer.

Figure 15:
View from the Gempen Mountain:
Autumn.

Through the course of the day and the year
we experience the reality of time.

Figure 16:
View from the Gempen Mountain:
Winter.

3. The context of life-processes in the landscape (context of life).

A third context is that which gives a landscape its individual character, and allows us to recognise a particular place whenever we return there. A landscape has a biography, whose traces can be found in many single details. When we attend to these with care, our resolves of will may become united with an ever richer experience of the landscape. We begin to notice how the formative tendencies of nature have collaborated with human activities – many of which may lie far in the past.

Sunny and shady, dry and moist habitats, speak such totally different languages. Animals join in, each in its own way. With the scenes composed of earth, water-surfaces and plant life, they take in a world belonging to themselves complementing their own nature.

Figure 17:
Aare reservoir near Koblenz, Switzerland, in Autumn.
Through the damming of the River Aare, deep and shallow areas were formed with a rich river bank vegetation. This in turn attracted a great variety of birds. A bird paradise came into being.

Figure 18:
Well-illuminated forest near Järna, Central Sweden.
From the anthill on the left, trails radiate in all directions. A badger-run leads across the clearing in a gentle curve. Thus animal life contributes to the mood of a place.

The different levels of context we have found in the landscape as a whole can be recognised no less in a single landscape feature, such as a tree.

The context of phenomena:
A tree always grows in harmony with its environment. The more or less clearly defined form stands in spatial relation to the overall illumination.
The context of transformation:
The vitality of the tree becomes manifest in the growth form of the branches.
* Rigid, upright growth gives the impression of youthful, energetic freshness.
* Irregular, angular branch-forms with slow growth at the tips give an impression of aging.
The sequence of annual growth-segments reflects the course of developments over the years.

The context of life
A rich, pictorial life history of a habitat can result when we proceed from the individual form of a tree to its development and transformations. The aims of the people who have played a part in this process are inseparable from the life-context.

Figure 19:
A young oak with its rigid, radial branches looks more like a wild cherry than our usual picture of a 'gnarled oak'. But irregularities are beginning to become visible.

Figure 20:
Free-standing oak, roughly 80 years old. It owes its characteristic form not to its young growth, but to a continual dying away of buds and twigs, followed by irregular new shooting.

Figure 21:
An oak several hundred years of age.
The whole crown has broken away,
and yet it continues to send forth
new shoots from the trunk and
branches.

The life history of a landscape, as read in the appearance of the oak – two examples

Figure 22:
An oak of more than 250 years near Gempen, Switzerland. This 'oak in the woods', surrounded by roughly 50 year-old pines and beeches, was formerly free-standing. The winter scene allows one to imagine something of the breadth of view one could have when standing under the full crown of the oak, at a time when it was already over 200 years old. Later, most of the branches forming the original crown died in the shade of the rising forest. The others grew upwards with the young trees.

Figure 23:
An oak in a young spruce plantation near Dornach, Switzerland.
The tree originally grew at the edge of a high forest. About twenty years ago the area was cleared and planted with spruce. A new crown is trying to form out of young shoots on the trunk; it barely reaches up to the level of the old, dying crown. On closer inspection we find an even older part: a mighty, decaying tree stump, from which the roughly 60 year old trunk visible today grew as an off shoot. Two further such trunks were cut away when the forest was cleared. The picture of a coppice with standards and occasional woodland pastures opens up as a still more distant perspective into the past.

In the characteristic contexts within the totality of nature – that of the present moment, that of rhythmic development, and that of the life history of a place – the elements earth, water and air work together with the warmth in specific ways. The entities which reveal this activity have a directly visible aspect as stones, plants, animals, man; others which have a part in the process are visible only indirectly in the relationships between the visible entities (elemental beings).

How can we see the elements and their activities?

Only insofar as we carry the particular qualities of the elements within us, can we find and recognise the pictorial expression of their activities without. When we look at a stone, we are impressed by its immutable shape. Our mental picture of this isolated form is more rigid than our perceptual picture, which is modified by our changing position and by the illumination.

Folded layers of stone suggest the picture of a fluid condition. The object has not changed, but the mode of observation has. Thus the plant kingdom, and each member within it, can be considered in the light of the elements, and how their different qualities become manifest.

The earth, with its topography and contours, provides the ground on which the plants can grow separately, and develop in different directions, according to their exposure. Our human experience keeps step with the resulting differentiation.

Figure 24:
View into the Tiefental near Dornach, which rises from west to east.
In such a valley with east-west orientation the horizon is narrow, and the gaze confined to an enclosed space. Agriculture is also limited within an area which is easily surveyed.
The isolation from the wider surroundings is stressed by the vegetation. Plants which are characteristic of cool, moist habitats in northerly regions grow on the southern slope of the valley; while the dry vegetation of southern regions appears on the northern slope.

Figure 25:
Landscape in the Rothaar Mountains in Southern Westphalia.
The converse is true for an east-west mountain ridge.
The view is open into the distance. And the plant world draws the mood of this breadth of view into our immediate proximity in the corresponding habitats of the northern and southern slopes. By contrast, flat landscape tends to monotony.

Water creates a space which moves and circulates within itself, and is divided from the wider surroundings by a mediating, harmonizing surface.

The formative development of the plant can be seen as an expression of these qualities. In the form of humidity, water represents the transition from earth to air. Under its sway, the plants swell in gentle, rounded forms; in arid conditions the forms become more differentiated.

Figure 26:
Water crowfoot on the open surface of a Finnish lake. The submerged leaves, divided into numerous fine, algae-like threads, testify to a flowing, dissolving quality prevailing beneath the surface. In contact with the air, the dissolving of the tips is arrested, and the leaf forms a rounded, gently lobed surface from within, giving expression to the quality of surface-formation through which the water divides itself from the air.

Of the four classical elements, air is the least accessible to direct sense-perception. It is there, and yet not there. The higher the vertical growth of the plants, the more specifically do their form and colour bring to expression their place in the habitat. These features are in turn subject to the illumination and changing seasons.

Warmth initiates and strengthens all the processes whose direction is predetermined by the interaction of the other elements. In plant development this means growth no less than differentiation and ripening.

Figure 27:
Interior of a damp, shady river bank woodland in view of a rather dry pasture with a rich variety of wildflowers. High-growing plants along the verge form a barrier to the light. Out in the open under full sunlight we can catch a glimpse of a rich display of colour.

Earlier cultures were directly aware of the 'genius loci', whose peculiarities and needs could not be ignored with impunity. Today the landscape, like the whole of nature, is regarded as something external to ourselves. We stand apart from it, as observers. Our feelings seem unrelated to the things without, and yet they arise as reactions to the outer world, and unite us with it.

Nevertheless, we have an overall impression which can be clarified in its spatial and temporal aspects in the way we have described.

By continually referring back to this overall impression, we can look at each stone, each plant, each animal, and attempt to find its place in the totality out of the specific quality of its appearance.

Figure 28:
Damp forest clearing with willows and alders in June.
In such a habitat – waterlogged and partly shady, but open in early spring – we would expect to meet the marsh marigold. And indeed, its leaves are in evidence; the peculiar character of the place seems to attract the plant.

Figure 29:
The same forest clearing with blossoming marsh marigold in May.
When the marsh marigold appears here in the spring, it gives pictorial expression to the dampness of the habitat through its soft, rounded leaves; the rich yellow of the flowers imparts a special character to the place. In this picture we experience a relationship which can only be brought to full clarity by the living activity of thinking. Yet this inner experience can nonetheless be recognised as an effective principle in the outer world of appearances.

Related plant species can characterise quite different localities, giving them a particular quality and expressiveness at certain times of the year:
Wood anemone and pasque flower, alpine squill and bluebell.

Figure 30:
Spring in a southern English woodland. Oak standards and coppiced hornbeam.
The form of the wood anemone, with its soft, finely-structured leaves, and delicate, gently swaying white to pink flowers, conjures up something of the mood of the damp spring forest, which is still open to the light.
For the wood anemone, the year is already finished in June. The deepening shade of the forest trees and the diminishing moisture bring about the plant's autumn. The delicate structures of the leaves and flowers quickly wilt. Only in the rhizome, which runs along just beneath the surface of the humus, is a bud preserved for next year. Over a number of years whole colonies can spread out from a single plant through underground growth, until a sizeable patch of forest floor is covered by an anemone carpet. The full reality of this habitat in all its complexity can, of course, only be touched on here. It came about through the activity of human beings, who have periodically coppiced the area right up until the present time.

Figure 31:
The pasque flower, by contrast, epitomizes something of the spring mood in dry, sunny habitats.

The radiant purple of the inner part of the flower is only fully visible in sunlight at the height of flowering. As long as the bell-like flowers are still hanging downwards, it is veiled by a coat of fine grey hairs which covers the whole plant. It obtains access to water through a deep, powerful tap-root. Where the pasque flower occurs in greater numbers, it does not spread itself out like a carpet, but rather tends to gather into isolated 'nests' formed by a single plant. During summer the plant retains numerous leaves, formed in rhythmical sequence.

Such distinctly characteristic plants are often used medicinally.

Figure 32:
Alpine squill in the transition from mixed lime forest to beech sedge forest beneath a white jurassic rock cliff in north-west Switzerland.
Slightly damp, highly porous limestone-rubble soil, rich in humus and nutrients.
At this time of year, in March, the trees are still leafless. The forest is still strongly illuminated, and receives a good deal of warmth from the bright, south-facing rock cliff. The green-flowered evergreen plants like the stinking hellebore and the spurge laurel are now past the peak of their blossoming. Now the most intensive colours of the year appear with the alpine squill and the spring snowflake in damper areas, and the cowslip in warmer places. On the rock we can discern the radiant pink blossoms of the delicate sand cress (Cardaminopsis arenosa).

Figure 33:
Bluebell forest in southern England
Here on the damp, partly sunlit floor of a scrubby woodland, the blue colour appears in much greater concentration, together with the green leaf-carpet and the first light-green foliage of the trees.
Alongside the bluebells appear the radiant blooms of yellow archangel; in an oak-hornbeam woodland on fresh, rich, acid clay soil.

The whole in the part

The same plant species grows differently according to its habitat; and different plants growing together in the same place show common features in form and colouration which can in themselves become objects of study. Each small plant, each branch of a tree blends into the whole, is in itself a picture of the totality belonging to a specific part of the earth.

Thus, each plant, each natural object becomes like a dewdrop in which the surrounding world is mirrored.

Figure 35:
Spear thistles in bright and shady surroundings on the edge of a pasture in central Sweden.

Figure 34:
Dewdrop on a lady's mantle leaf. In each dewdrop the world is mirrored from the perspective of a single point.

To experience the plant like a dewdrop which mirrors its surroundings, means to become intuitively aware of a great diversity of present and past relationships, as they are manifested in its outer appearance. Or, in more general terms, to apprehend something of the totality of its relationship to the interplay of cosmic and earthly forces.

In this way the creative idea of the whole place or landscape in question can be experienced from innumerable aspects.
By bringing more and more of these aspects together, and uniting them, as it were, into a single, artistic composition, we can begin to fill out with substance and content what we have grasped of this underlying idea.
We gradually become participants in a creative process.

Figure 36:
Moist habitat in central Sweden. The marsh thistle is a close relative of the spear thistle (cf Fig. 35). Although it favours the damper conditions, it nevertheless develops much sparser and more finely divided leaves towards the top of the stem. This apparent anomaly is due to the poor, acid soil. The strong reddish tinge on the ridges of the stems harmonizes with the similarly-coloured stems of wood avens and meadowsweet which are growing around it.

A balanced and healthy relationship can only be fostered where the individual is able to enter into a dialogue with the spiritual forces at work in nature, and to make this the basis of his actions. This will obviously need to include the choice of appropriate forms of technology. Vigilance will be needed to assure that this takes place in freedom and not under the influence of social or technical constraints.

At the same time, he will have to seek his place within the social framework. The resulting human contacts may well pave the way for the other to recognise and accept something of the spiritual realities he is attempting to work with. In this way, understanding may be forthcoming, even when outer criteria would seem to make his actions appear irrational and incomprehensible.

Thus, a free space is created for transformation. Circumstantial constraints gradually lose their power.

As new interests arise, a new life style will develop, providing the individual with a new and secure foundation from which he can turn his thoughts and deeds to all that makes life worth living. In consequence, his criteria for dealing with material needs also change.

On closer consideration, whenever a true partnership is established between man and nature, opportunity is given for the development of what is individual. This is the essential nature of what we could call the idea of a landscape.

Human interference with the course of nature usually has the effect of destroying or suppressing what is already present and replacing it with something new. By adopting such a policy, man relies on the response of the plant kingdom. But what is the ultimate goal of his intervention?

Where man so collaborates with nature that its qualities and forces are allowed to develop, beautiful landscapes come about.

Figure 37:
Golf Course on the edge of the Ashdown Forest in Sussex. The forest here has been cleared, and the grass is closely mown. Under the influence of England's maritime climate, such artificial clearings attract special plants, which in turn adorn the landscape with its own expressive beauty.

Where purely commercial thinking prevails, the landscape becomes monotonous.

Figure 38:
Modern sheep pastures in New Zealand.
The virgin forest was destroyed, and in its place alien grasses and herbs were planted. Even the trees which make up the wind-breaking hedges are imported.

Figure 39:
The mysterious, yet inhospitable Kauri Forest in New Zealand, parts of which were cleared to make way for sheep pastures.

Which forces are most widely utilized today in farming and forestry?

In habitats which have been allowed to develop undisturbed over decades or centuries, the soils take on an organised structure, and the plant communities become quite varied.

In a forest, for instance, a kind of dynamic equilibrium is established over a longer period of time among a variety of plants which mutually complement each other. Depending on weather conditions, different species will predominate in each particular year. Growth is limited.

We find a living stability amid rich differentiation, while productivity remains relatively low.

Figure 40:
Natural Finnish spruce forest in July.
In the north, as at high altitudes, small plants and herbs display simple forms and often bear white flowers. Nonetheless, there is considerable variety here (cf. Fig. 5).
'Spring' comes very late.

If the trees are felled, luxuriant, disorderly growth will ensue. Nature needs much time to restore harmony to such an environment. But does it still have the necessary vitality, when left entirely to itself?

Figure 41:
Recent clearing with transition to the forest. New offshoots from the stumps produce large, mildewed leaves. Plants which previously developed sparsely in the forest shade spread out luxuriantly.

The consequences are similar when a meadow is ploughed. The individual plants develop side by side in their different ways. Each is an image of its particular surroundings. Only a few attain to their optimal development, which in all events remains far behind the maximum possible size and overall mass. The activities of slugs, aphids and fungi remain inconspicuous, as they are directed at the weak or wilting plants.

After the field is ploughed, the perennial plants either disappear into the background, or begin singly to luxuriate. Large biennial and fast growing annual plants dominate the field and create the impression of total chaos. The tendency to grow together into a differentiated unity can hardly come into its own.

Uncontrolled, luxuriant growth provides an entry to pests.

Figure 42:
A permanent pasture on the edge of a forest in central Sweden. The rich variety of flowers is the result of many centuries of regular care through mowing and pasturing.

It thus becomes increasingly clear that man's task is not merely to increase the productivity of nature, as for instance through the use of highly soluble fertilizers, but rather to assist in establishing order and harmony. To be sure, monocultures and the use of poisons do bring about a certain order, but one which is imposed on nature from without. For this reason, the cultivation of land and forest has tended to become ever more one-sided.

Here we can gradually learn to recognise the difference:
* between healthy ripening, in which the process of wilting/dying as a limiting and cosmically formative principle works together antagonistically with the process of life and growth,
* and forces which bring death in their wake, and in the long run mean total annihilation.

We need the vegetative, luxuriating forces for the cultivation of the land, and we enhance them through the use of fertilizers. But in so doing, we suppress the cosmic, differentiating forces which otherwise bring about the natural, healthy ripening process.

This imbalance requires adjustment in two directions:
* The differentiation of the landscape as a living whole, in which natural preserves (forests, hedges, marshland and the like) are retained alongside the cultivated areas. Here the cosmic forces of differentiation can develop and enhance the health and quality of life in the wider surroundings.
* Agriculture. While the stimulation of growth through the fluid medium (highly soluble fertilizers) merely increase mass and water content, the environment of the soil through carefully prepared organic fertilizers, and the development of humus appropriate to the site enhances the plants' direct receptivity for the cosmic forces. These in turn contribute to the maintenance of a healthy constitution and to the formation of genuine nutrients. In other words, they work towards a kind of inner-differentiation.

Figure 43:
Shell limestone landscape near Bad
Mergentheim, central Germany

Individualizing work with the land.

We have seen how our modern way of life suppresses the very forces which would work to create a healthy and harmonious whole.

In a life situation it is not enough merely to ask whether something is correct and feasible (out of a sense for the facts). Nor will it suffice to direct our attention to the whole web of relationships out of which it arises (motivated by a sense of beauty). For it is no less vital to discover where and how it will affect the rest of nature and society.

Figure 44:
Such a path, cutting into this loess woodland like a dark tunnel, can open up new channels of interest for the whole varied life community around it. One can hope, in turn, that the clearing of the fields will stop short here, in view of the rich variety of life which a habitat of this kind can radiate into the wider surroundings.

Figure 45:
Agriculture which works towards a harmonious balance of pastures, meadows and fields; of livestock-raising and field cultivation, with respect to each other and to the local and climatic conditions. The forest surrounds and protects. The low woodland at the forest's edge provides a link with the farmland through hedges.

For this it is necessary that the areas of land worked by man should be small enough to allow a survey of the whole. Only thus can they be managed and fashioned in an individual way.

Fruitful beginnings in this direction can be found in the worldwide biodynamic agricultural movement, and in many 'appropriate technologies' developing under the influence of new social forms.

But the renewal of responsibility in the sphere of land-use and technology is only thinkable in conjunction with a concerted schooling of a sense for what one's own activities will mean for the rest of the world (cf. p.30).

Every single object has its place in a larger totality. We tend to take this totality for granted, but through it we too can find our connection with the world at large.

We can enhance this connection by consciously schooling ourselves in the observation of natural habitats which are still intact. In this way we can gradually develop a sensitivity to the cosmic relationships at work there. Such relationships, which are themselves invisible to the outer senses, yet nonetheless essential to the health of their habitat, have widely fallen into disorder. Indeed, their very existence is threatened by our habitual attitude to life, and by the destructive, isolating effects of technology.

It would be a serious mistake to assume that all this concerns only a few farmers, foresters and landscape architects. What is needed is a radical new step in consciousness, which is ultimately the responsibility of each individual. No one can take this step on behalf of another. And this means a total transformation of our relationship to nature. Life will become more difficult, but at the same time richer.

The kind of schooling we have described also enhances moral consciousness. This means:
*   The ability to bring one's own actions into harmony with nature and society alike.
*   Finding a meaningful and appropriate place for modern technology within a healthy life-context.

There are countless opportunities to put this into practice – for instance, in planting a tree, to consider:
*   whether the species is appropriate to the habitat,
*   whether it was raised under conditions which correspond to those prevailing here,
*   whether it is the right time for planting?

But equally, when we purchase something, to be interested in:
*   where it comes from,
*   what landscape and what people had a part in its production,
*   what should we do with the waste?

If we have a garden, and save organic waste for the compost, we will naturally make sure that it contains nothing unperishable. But what should we do with the rest? We will soon realize that not all of our practices are ideal, but we have nonetheless taken a first step in assuming responsibility towards our environment.

Figure 46:
Someone planting a tree with foresight.

The dying of forests, the extinction of countless species of plants and animals, and the appearance of pests and diseases on an epidemic scale – all these are but symptoms of a wider death process which seems to be spreading inexorably on many fronts.

It is an image of modern man's way of living and thinking.

An attitude to life which is the outgrowth of a one-sided direction of science stands behind the modern practices of agriculture and forestry. Luxuriant vegetative growth alone is encouraged; chemicals and deadly poisons are used to steer and protect this imbalanced process. Through industry and traffic, this attitude has worked to create a subnatural realm which is cut off from the living context of nature. Fragmentation and isolation are its inevitable consequences. The final result can only be destructive to life. And therefore the wellspring of formative cosmic forces, which can only adequately be absorbed by a rich variety of life-forms, is in ever more critical danger of being exhausted.

But the pollution of the environment through industry, household waste, and traffic is also a testimony of the fact that we have failed to come to terms with something in ourselves.

Under the sway of our wishes, we are all too prone to embark on activities which we cannot adequately transform through our inner life of soul and spirit.

This too seems inexorable. Genuine responsibility can arise only out of freedom.

For by following outer laws alone, one relinquishes responsibility to the law-givers. The divergence of natural law and moral experience was necessary for the development of freedom.

But responsibility can also only arise where one is able to find a connection to a totality which can be surveyed and comprehended in all its aspects. What is most needed therefore is a path of knowledge which gives us access to this totality.

In ecology, great emphasis is placed on the interweaving of concepts into a totality, or 'network'. But every connection between two things creates new concepts, which in turn must be woven into the tapestry. In the end this leads nowhere. It is not through conceptual combinations of outer things that the totality can be grasped.

By contrast, the image of the dewdrop can help us: The totality can only be realized through an inner act of knowledge. It can be experienced in each single sense-phenomenon at the moment we become aware of what we are seeking – in this case the individuality or essence of a landscape or habitat.

The facts of the world are not pre-established in just the form in which we later apprehend them. What we recognise as a fact depends on our attitude to the phenomena.

In our conscious intercourse with nature, we learn to bridge over the gap between our inner soul-life and the outer world in a spiritual reality which penetrates them both. We learn to think in terms of the whole, and thereby to give a new direction to our actions.

The cultivation of a sense for beauty and the loving devotion to the action on which we are resolved, make this action satisfying in the very doing, quite independently of possible material benefits. Actions thus motivated can help us to heal what is sick in nature.

The more we allow our interests to be dictated by the outer world, and the more we compel this world to serve our personal and social needs (eg. through machines), the more the economy will grow – but we will become lonely, driven by constraints, and the forests will die.

The more we direct our interests to what is spiritual in the world – that is, to what is akin to our own being – the more the people and things around us will grow in meaning and importance to us.

Here lie the seeds for a new culture. Technology will be used only insofar as it can provide us with a free space to become active and creative ourselves. Values will be transformed right into the economic sphere. The air will become cleaner again.

The joy which these changes will awaken in people will play an important part in helping to further the new, healthy methods of forestry and agriculture.

The life of nature in a particular locality can begin to take on the form of a fluid pictorial composition, in which changing conditions find their expression. The special relationships of warmth, light, moisture and soil; the seasons and their effects on the life communities – all find their place in the whole. The contours of the earth provide the groundwork; the colours fluctuate with the growing and wilting of the plants. The animals make but fleeting additions to the colour. But they add another level, which finds its expression in the world of tone. Thus the mood of the open fields seems to ring out from on high in the early summer song of the meadow lark. The robin's notes convey something of the sensation of morning or evening in a rather shady part of the woods, while the resounding song of the wren seems to suggest the enlivening sound of a bubbling stream hidden from sight in the bushes.

But there is more to it than this. The animals perceive the scenes composed of earth, rocks, water-surfaces and plants. They are attracted or repulsed by what they meet. They can also play a part in fashioning what appeals most directly to our feelings in a natural scene. This may take the form of a bird choosing a particular spot to build its nest. It can speak through the manifold relationships of insects to flowers, or even in the hidden, structure-giving work of earthworms and other small animals in the soil. Each activity and movement which has its source in the inner life of the animal world plays its part in stressing and 'making audible' the mood of a place.

This is not to say that a soul experience should be ascribed to the place itself; but every form, every colour composition has its own objective form of expression, and this awakens a certain inner response. Each percept is like a question or challenge which calls for an answer from within; and this is what gives rise to a sense of

balance and satisfaction in our experience of the world around us. Only when the sense-perceptions are blended together into a pictorial whole do they attain to their true significance. Each kind of animal is distinguished by the particular aspect of the world which it perceives. And therefore each takes in the world through different kinds of scenes and pictures.

One might be tempted to think that the same is true for the plants, when they are growing in a habitat specially suited to their life requirements. On closer observation, however, one finds that the plant reacts quite differently to its surroundings. It does not display a special 'interest' in something visible, audible or otherwise perceptible. It does not focus its attention, like a butterfly on certain flower-forms. Nor does it react like a nestling to a certain beak-form and colour of the parent bird. The only thing which might be regarded as comparable is the tendency of the plant to follow certain directions of development when exposed to light, moisture, warmth or gravity. But just this kind of process shows how differently the plant relates to the world around it. For what is light, or brightness, what are warmth, gravity or moisture? In themselves they are not accessible to the senses; it is their effects which we perceive.

Like the animal, we can perceive these effects through the medium of the concrete, material world. The plant does not perceive in this way. It absorbs these effects directly and develops accordingly. A pansy seed will grow and develop to the characteristic form of the species. The idea which we at first connected with the seed has become pictorial expression, an outer reality. It assumes a certain orientation in space and in the field of gravity; the forms of the leaves and the colours of the flowers are more richly differentiated in sunlight than in the shade; growth is more luxuriant when the soil is moist and rich in humus and minerals, than when it is dry and poor.

As an autonomous being, the animal responds to the forms of its perceptual world (which may contain very few specific elements) with independent movements, which have their source in an inner life. The plant absorbs influences from its environment and translates them into forms. The picture of the species, which reaches its highest expression in the flower, develops through a sequence of forms. These are brought to a standstill and then fade; new forms grow to replace them. The process of transformation, which leads dynamically from form to form, is not the expression of an independent movement. Thus the plant is eternally in a state of flux, progressing through a sequence of forms, and as an organism never attaining to the autonomy of the animal. Through the plants, pictures are engendered which awaken experiences. The plants do not have feelings, but they permeate space, as it were, with qualities of soul, to which beings capable of feeling can find a relation.

But the plants do not stand in isolation. Each plant seed is attuned to a particular composition of environmental influences, under which it can develop in an optimal way. If this attunement is lacking in a particular area, if a place is too shady or damp, or if the rhythms of the day and the year are not suitable, then the species will not be able to develop, even if its seeds are present, whereas other species will flourish. In this way a plant community will develop out of the particular conditions prevailing in a habitat. One plant within this community will create the conditions under which another can thrive. A living totality develops in which, as it were, the different voices of the composition can be discerned. Woody plants, together with the geological 'groundwork', constitute the more permanent elements, while the annuals are more fluid and shifting.

In this sense it is justifiable to speak of the 'mood' of a place. The overall situation of a moist forest clearing, through which a stream is flowing, creates a suitable habitat for the marsh marigold. The mood which characterizes the place as a whole attains a special nuance through the roundish, juicy leaves and the deep-yellow flowers in springtime. But such nuances are transient, and subject to continual change during the course of the seasons. Thus, a place can be said to have a certain physiognomy and gesture.

What thus begins with the plants is further enlivened and differentiated in the most manifold ways by the animals. We have to do with a life-context which in each case can be grasped spatially as a seasonally changing picture; and which above and beyond this, undergoes a kind of individual biographical or historical development (eg. in the growth-forms of trees). This process of individuation can only be found to a limited extent in 'untouched' nature.

Thus far we have not spoken either of 'landscape' or of man. Both belong together. One can only meaningfully speak of a landscape if the life-context as the mood of a place is mirrored in human consciousness, and is more or less intentionally transformed by human aims.

In the case of the animal, one can only speak of its world, of which it is an integral part. In accordance with its psychic organisation, the animal, in its dream-like consciousness, weaves each new sense-impression into the totality of its world, and acts in response to these impressions. Thus a warbler, which orientates its nocturnal migratory flight in spring and autumn by the stars, can be deceived by the light patterns of a planetarium. Man can question the truth of his sense-impressions, and enquire whether the starry firmament he sees is real or illusory; moreover, only he can create such artificial models. He can distinguish the essential from the non-essential.

The life-context of a place with which one is connected as knower or doer can also be comprehended through the same critical faculty. But it depends on the observer, how broadly he is able to see this context when he refers to it as a landscape. There is more involved than just something 'out there', or something which only lives in the observer. An inner and outer reality meet in what we would call the sense for beauty. This transcends the individual details, and directs attention to what is more or less recognisable as an inner unity in the scene. What is meant is a spiritual unity which lives in the effective relationships between natural processes and human intentions.

In ecology today, one often encounters the view that we only need study the relationships between the minerals, plants, animals and man in order to come at last to a unity through the 'interweaving of concepts'. But this is not possible unless it is accompanied by the quest for unity on an ideal plane, for only here is the unity a full reality. Obviously, we must be able to take for granted that such a unity can indeed be found. Our first general impression of the quality of a landscape is like a first intimation of the underlying idea. As such, it can serve as a guideline for further study. But because this impression tends to be rather vague, it is easily forgotten, or rejected as a useful criterion.

For various reasons, our description of the cognitive process is likely to be received with scepticism by those grounded in an ordinary scientific training. They will tend to be wary of subjective elements entering unconsciously into this process. They will all too easily misunderstand what we mean by 'idea', interpreting it as a fixed preconception which arises in conjunction with the initial impression we have spoken of. If this were true, their objections would be justified, for we would be imposing a subjective mental picture on the context of perception. What is meant is something quite different. The idea of a landscape is taken to be an effective reality at work in the outer world. Stage by stage we can begin to apprehend something of this reality out of the confidence that it is possible in principle to find the effective, spiritual context of a chosen part of the surroundings, which can in turn appear as an aspect of the world-unity. What we have already grasped of this idea assumes the role of an inner organ of perception, helping us at each stage to raise further aspects to consciousness.

Not only the processes of nature, but also the quality of thinking and experience of the people connected with a place find expression in the landscape scenery. In earlier times little thought was given to these things. Beautiful and satisfying scenes arose, as it were, spontaneously through an interaction with nature which was grounded in a healthy life of feeling.

With the development of a new quality of thinking in the dawn of the scientific age, man began to lose his instinctive connection with nature. His technological ideas led to the construction of machines which liberated him from manual labour. This process began in the cities, where the effects on social life were at first more drastic. In a certain respect it meant an emancipation from outer constraints, but at the same time from certain obligations which had previously been taken for granted. Gradually, the new technological ideas began to find application in the sphere of natural life processes. Nature was degraded to a source of production, comparable to a machine of human construction. It answered with unexpected 'side-effects' parallel to those which had arisen earlier in the social sphere. There is no need to describe the consequences here in detail. Nor to show how these things developed with a certain logical consistency in the course of human history.

But it is obvious how this quality of thinking is increasingly reflected by the landscape in the form of monotonous agrarian districts, treeless wherever possible, and monocultures with their dead-straight boundaries. The want of any feeling of obligation also comes to expression in the fact that every remnant of local history fades out of the landscape scenery. Everything becomes impersonal.

Now it is a remarkable fact: "Whenever through the advances of culture and civilisation man loses touch with nature as the foundation of his life, he aspires to become consciously aware of its significance. As a rule, the feeling for nature and landscape is more strongly developed, the more nature is endangered. But the development of a feeling for nature is by no means a guarantee for an appropriate treatment of nature and landscape. It is rather a signal for the contrary. History is too full of examples of over-exploitation and destruction of intact landscapes to give support to the theory that the development of a feeling for nature must automatically constitute the necessary counter-movement" (W. Nohl, Landschaft und Stadt, 1982). And yet, only a conscious approach to the idea of a landscape can provide the starting-point for renewal.

The painters of the Romantic movement, which coincided with the beginnings of the technological age, were the first consciously to discover and record the landscapes which had developed in Europe since medieval times. At the same time, ideas on landscape architecture arose in Central Europe which strove to synthesise the Romance and English schools. Important impulses in this direction are to be found in the writings of Prince von Pückler-Muskau ("Hints on Landscape-Gardening", 1834).

Prince von Pückler is able to look back on the prototypes of the Italian Renaissance and the French Baroque, with their ideas derived entirely out of the human world, and thence imposed upon nature. He comes to value the English style, which prefers to allow nature free sway, with support and encouragement from man.

For his own part, he attempts to arrive at an independent approach. He is filled with the longing to awaken deeper levels of human experience in communion with nature. In this sense he is a true representative of the Romantic Age. Like other great personalities of the period, he directs his gaze to the working of the spirit in the world, to the ideal sphere, and does not dwell in the superficial realm of playful fancy. His goal is a 'region ennobled by art', 'nature idealized through art'. His starting-point is not merely a planning-scheme; rather he attempts to become conscious of his intercourse with the idea, as well as its nature and origin: "A great landscape garden in my sense must be based on a fundamental idea".

"Let me be properly understood. I repeat, a fundamental idea should underlie the whole; no confused, muddled haphazard work should be allowed, but the leading, perfecting thought should be recognised in every detail. This thought may have its origin in the particular striving of the artist, in the circumstances of his life, or in the locality which he has to deal with – but by no means would I expect that the whole plan should be precisely drawn out in every detail from the outset, and then strictly followed. In a certain respect I would even recommend the contrary procedure; for even though the main features of the whole may be pre-established with the idea, the artist should be under no compulsion, and free to follow his artistic inclinations at each moment. In this way he will constantly make new discoveries, and will continually study his material during the process of creative work. He will observe untouched nature as he finds it both within and without the realm of his little creation under all conditions of illumination, for light is one of his chief aesthetic resources. He will explore the causes and effects, and in accordance with his findings will specify his earlier, single thoughts to fit the details; or alternatively, will abandon them in part, should he later attain to a better insight."

How such a method can be applied in a concrete situation may become clearer from Prince von Pückler's later description of his own special planning idea for Muskau in Lower Silesia:

"Having adequately familiarized myself with the area, and with the prospects for achieving my plans, I decided that in addition to the existing garden, the whole river area should be incorporated in the parkland. This would include the neighbouring plateaus and hills, pheasant-preserves, meadows, farmsteads, mills, alum mines, etc. from the last ravines of the southward sloping mountain ridge, as far as the villages Köbeln and Braunsdorf on the northern side: altogether, nearly 3000 acres of land. Moreover, by including the slope which extends beyond the city, along with a part of the village Berg which is situated on it, the city would be so embraced by the park that in future, together with its surrounding meadows, it should virtually constitute a part of the same. As it is a mediatized city, hitherto subject to my rule, its inclusion in the projected whole acquired historical significance; for the main idea upon which I based the whole plan was none other than the creation of a judicious image of the life of our family, or of the national aristocracy, as it has developed pre-eminently in this place. It was my aim to make this image visible in such a way that this idea would, so to speak, arise of itself in the feelings of the beholder. To achieve this, it was necessary to to make use of what was already present; to enrich and enhance it along similar lines, and yet never to violate the locality and its history. Many an ultra-liberal spirit would no doubt smile over such an idea; but every form of human development deserves in its own way to be commemorated, and just because the form now under discussion may well be nearing its end, it is beginning to acquire a general, poetical and romantic interest which can hardly be equalled by factories, machines and even constitutions."

As an idea in the higher sense, which underlies all garden and landscape art, he sees "... to create out of the totality of the natural landscape a realm of nature on a smaller scale as a poetical ideal. This is the same idea which underlies every true work of art and which recognises in man himself a microcosm, a universe in miniature."

That this idea could be realised in Muskau just as Prince von Pückler describes in detail, is a consequence of the dying traditions which still lived for human beings of that period. More important is his pioneer deed of raising such an idea to consciousness in a Goethean sense. The creative act should neither consist in imposing on nature a preconceived plan, nor in merely allowing things to run their course in a haphazard way. For even by just 'muddling along', man would still imprint unconsciously on nature what lives within him.

The prophetic nature of his procedure lies in the striving to ascend to the idea in its purity, and from thence to attempt no more than to make visible human ideals, as manifested in the quality of life in a particular community, in relation to a certain region and historical period.

These beginnings found no continuation in the period which followed. From the middle of the last century onwards, human attention turned predominantly to material and industrial culture. In the course of time, this led to an ever-increasing material knowledge in the realm of garden architecture, stimulated by science and pragmatic gardening techniques (systematic collection of plants from all parts of the world, and ecology). But the ideal side was lost sight of. Only after the recent awakening to the consequences of a one-sided attitude to nature, were the romantic ideals remembered by the circles responsible for conservation and landscape-planning; but no change was forthcoming in the sphere of human aims, which had come to be dominated by material considerations.

If we are able to enter into the spirit of Prince von Pückler's ideals, his direction of work can certainly be carried on into the future. He would direct our attention to the idea which underlies the relationship between man and nature. The ideal which he held up for a work of art points in a direction which can help us to work against the destruction and loss of variety in nature, even though he himself, at the dawn of the industrial age, could not yet clearly foresee its consequences.

Under von Pückler's idea of a landscape we thus understand the whole context of relationships which connects the people of a certain region with the life conditions of nature. It can be more or less outwardly manifest, and can be consciously taken hold of and further developed by the artistic sense of man. What lives in man becomes visible in the landscape. This does not need to be limited to traditional forms of life. It does not need to relate only to the past. If in the course of human thought and action attention is paid to the idea at work in a landscape or estate, a way is laid open to re-enliven the forces of nature. What takes place among the visible entities (the soil, the rocks, the water, the plants, animals and man) is under the sway of the idea. It can take form more or less clearly. By becoming interested in how this can take place, we can allow the idea to speak to us all the more clearly, so that we in turn can help its realization.

We are thus concerned not with an idea which is imposed from without, but rather with one which is implicit in the life-context of human-beings with a locality, and which seeks to find a fuller expression. It can only be apprehended and brought to life gradually, through active experience. This means that not only thought-out forms, or the goals and requirements of production should be brought consciously to bear on the fashioning of the landscape, but also the life requirements of the entities around us and the cultural striving of man.

This attitude underlies the indications for agricultural renewal by Rudolf Steiner (1924) which led to the development of bio-dynamic farming and gardening. He shows how man, with his justifiable material and spiritual interests, can become a meaningful part of the natural processes in a locality. Farming thus finds a healthy connection to shaping and caring for the landscape. In this way an enhancement is achieved of the spiritual substance out of which a farm can be shaped to an individual, fruitful totality.

A landscape becomes all the more individual, and consequently attractive, the more its life is penetrated by human influence, without incurring a loss of natural processes and functions. The individual quality – and with it the effective idea of the landscape – is extinguished if a preconceived idea is transferred from the drawing board to nature. Procedures of this kind erase the past while attempting to fix the future. The development of an individual character can be encouraged when one seeks its features in the present landscape scene and tries to discover what would like to happen. In this way one is not working with a finished picture, but one begins to participate in the creative working of the idea in the context of life.

Science and ethics

The problem of ethics and social responsibility arises unavoidably from the irresponsible applications of scientific research which we encounter on all sides. Indeed, we may well ask whether ethical criteria are possible at all in a scientific community which wishes to remain unencumbered by human values in its approach to learning. In former times man experienced the coherence of natural phenomena as 'cosmos', as the order and beauty of the world. This was intimately bound to his experience of divinity in the world-order. He strove to shape his life in harmony with this ordering principle.

But today the man of science feels himself to be an external observer. He discovers universally applicable laws in nature, which he has learned to manipulate, making them the basis for technology. Whether or not he allows ethical considerations to play a role is his own private concern. Inner ethical experience is obviously unrelated to external natural law. The student is expected to learn these lawful relationships in abstract form, in order to gain a survey of the forces active in his field of work. The resultant technology is restricted only to what is practically feasible, and its orientation is dictated by desires and economic interests. A sense of responsibility, either to the content of knowledge or to its technical applications, cannot develop in this process.

Responsibility can only arise out of freedom, for by obeying outer laws alone, one relinquishes responsibility to the law-givers. The divergence of moral experience and natural law was a prerequisite for the development of freedom. But responsibility demands that we find our place in freedom in a natural or social context. This does not occur in a science that is transmitted only as factual knowledge. Conversely, we can only develop responsibility in an area which we fully survey.

How can scientific methods and procedures be modified to allow the principle of freedom its due, while yet achieving a survey that does equal justice to scientific fact and its human context? How can we gain the capacity to recognise the comprehensive relationships in a given instance and adapt our actions accordingly?

The ideal of a science free of value judgments is only valid in countering the danger that facts might be appraised solely in terms of social or personal advantage. But the ensuing scientific attitude has introduced into nature a 'reality' which is in principle alien to all value judgment. Every natural occurrence is thus interpreted as the effect of outer forces. The methods of mechanical science are taken as exemplary in establishing causal factors.

In reality, however, each unfolding situation belongs to a wider context. Apart from the mechanical factors involved, it is implicitly assumed to belong to the whole as well. This even applies to the realm of physics, although the context here (as universe) is so comprehensive that it escapes our notice. And yet we take it as a matter of course that such a context exists, in which we participate in the very least through the power of thought which resides within us. Otherwise, why should we trouble to understand anything? It may be more apparent in the organic realms of plant, animal and man, that chemical and mechanical processes subserve and obey a whole of a higher order. Here it is only the outer appearance of a living organism that can be grasped by mechanical, physiological thinking, not its essence. If we inject a substance into a plant in a certain way, then after a time we see perhaps that it produces blossoms, which it would not have done otherwise under the given circumstances. It is not the added substance that forms the blossoms; these emerge rather from the total context which we call 'plant'. The substance merely provides a condition – perhaps a decisive one. We confuse condition and cause all too easily.

In life situations, we are dealing not only with mechanisms derived from external forces, or with given organic structures, but also with such adjustments to natural processes as have been introduced by man, in his effort to place outer facts into effective relationships. Ethics and responsibility depend on the way he goes about this.

If he pays heed only to mechanical factors, then unpredictable side-effects are inevitable. If, however, he takes a comprehensive view, and tries to understand their relationship to the creative forces in nature (in our previous example, the plant – in Goethe's sense – as an idea producing effects in real life), he can integrate his own action more and more into the appropriate spiritual context. In this sense his action can become moral.

If as scientists we turn our attention not only to results, but also to our own activity, we will soon note three points at which an objective evaluation might ideally be introduced:
1. as we search for the meaning or the intrinsic value of the fact in question;
2. as we consider the qualitative significance of this fact to the overall context in which it stands;
3. as we reflect on the direction we give, or the meaning we assign, to an action, in the light of its natural or human setting.

On all these levels of evaluation we must accept responsibility. It therefore becomes necessary to embark on a concerted schooling of our consciousness of responsibility. We will need to concentrate on three corresponding areas:

1. Evaluating the facts which we seek to understand. If we are not concerned only with what we know, but with how our knowledge arises, we have already attained a vantage-point from which we can begin to assume responsibility for the things outside us. The validity of a scientific claim is normally tested by a method that follows from outer axiomatic assumptions. We follow in thought the steps the experimenter took and experience their inherent logic in the manner of a mathematical proof. By thus establishing the truth or untruth of the claim, we have assumed responsibility, we have united ourselves with it inwardly through our own conceptual activity.

Goethe laid stricter demands on the scientific observer than is usual. He demanded that the data for judgment be drawn from the phenomenal sphere, not from quantities defined by an a priori system. The latter procedure would seek other explanations for a particular phenomenon than those which directly follow out of the field of observation, such as the movement of particles as the 'cause' of the warmth we perceive. Moreover, the notion of moving particles offers no adequate causal basis from which we might derive or 'explain' warmth. If we encounter moving particles in our experiments with heat, we have to do with parallel observations. We may arrive at necessary conditions but not at insight into causes. A method that maintains responsibility will restrain the thinking from taking recourse to abstractions that transcend the given facts. It is moral, insofar as we take pains to restrict ourselves to observing facts and thinking through their relationships, without deviating from them arbitrarily.

2. Holistic knowledge. The second form of responsibility – directed towards the context from which our knowledge is drawn – assumes a holistic mode of cognition. To be sure, one often speaks of holistic knowledge, but from the usual standpoint of the external observer it is impossible to achieve. For a fact we confront only from without cannot constitute a whole; it can be no more than a part. Goethe, who was equally at home in the arts and the sciences, has shown how a natural phenomenon may be investigated in such a variety of ways that something of its essential nature can become directly accessible to outer observation.

He has demonstrated this particularly clearly in his colour and plant studies. (See the section on "Goethe's Scientific Method – A Path to Holistic Knowledge" p.22).

The comprehensive laws which can be found through this mode of observation include more than just the phenomena before us. They enable us to experience in thought a great variety of possible but unrealized forms in the plant species we are considering. We have illustrated this through the example of a series of field poppy leaves (see pp.22-23). At the same time they allow us to survey other contexts in which the plant may be viewed. Here too, we must be mindful of the mode of observation we are employing. We must know two things: first, whether the relationship we have established through thinking truly follows from the object under observation; and then what sort of answer may be expected from a particular kind of question? The following examples may serve to demonstrate how the nature of our approach, or the form of our questions will determine the relationships that come to light:

* We can ask how much water a plant will transpire under certain conditions. Then we are interested not in the plant itself, but in its performance.

* We can try to understand the forms of its leaves, as we have done in the example of the field poppy. Here we are applying a mobile faculty of pictorial thinking to investigate a formative process.

* We can study different plant species, learning to distinguish in each case an unmistakable gesture which underlies the most manifold variety of forms. In Figure 9 (p.24) we can recognise a similar law of transformation at work as in the leaf series of the field poppy. In spite of its variable forms, a specific expressive quality is inherent in each leaf, making it impossible for us to confuse leaves of the two species. The specific quality becomes still clearer, of course, in the flower, but it nonetheless penetrates each formative phase in the plant's development.

* A quite different point of view concerns the plant's relation to its environment. Seen in this context, the plant will furnish a picture of light, soil and seasonal conditions in its surroundings. In fact, it would not be too much to say that only through the plants do the environmental conditions and their influences come fully to visibility.

The transmission of knowledge of this kind is only possible when the appropriate quality of experience is implicit in the way the facts are presented.

3. Creating a clear basis for action. In a life situation it is not enough merely to ask whether something is correct and feasible (out of a sense for the facts). Nor will it suffice to direct our attention to the whole web of relationships out of which it arises (motivated by a sense of beauty). For it is no less important to discover where and how it affects the rest of nature and society. This

requires a moral awareness of the indwelling tendencies of an action to begin with, and then an intimate acquaintance with the nature of the beings it will affect. Here again, the problem arises of establishing clarity on a human – not only a material – level. Only in this way can adequate criteria be won for evaluating the likely effects of an action in the future.

It is therefore necessary to fashion the areas in which human beings should work in such a way that they can be easily and clearly surveyed. Beginnings in this direction can be found in the world-wide bio-dynamic agricultural movement, and in 'appropriate technologies', developing under the influence of new social forms.

In bio-dynamic agriculture, Rudolf Steiner has shown how we can farm in conscious harmony with nature. Increased fertility is achieved through the 'process of individuation', which works to bring fields, meadows, pastures, woods, hedges, livestock and farming into mutual attunement, in harmony with the local and climatic conditions. This reduces to a minimum the expenditure of energy and materials from outside the farm.

Within such a holistic framework responsible action can unfold, since we are able to survey the organic cycles within the farm. The fruitfulness of this method has been widely demonstrated, both under European conditions, and under those of countries in the so-called 'third world'.

Summary

Thorough, practical acquaintance with nature can help us bridge the gap between the inner world of feeling and the outer sense-world, by means of the spirit which permeates both. We learn to think in terms of the whole, and let this become the basis for our actions. We gain the ability to bring our actions into harmony with both nature and society, and to integrate the newly created world of technology meaningfully into the context of our lives.
We all make use of the capacity which Goethe cultivated in his scientific work – that of comprehending life relationships through the active power of thought. It can be raised to consciousness, be made a matter of experience and developed further. From this starting point, Rudolf Steiner further refined Goethe's path of knowledge.

A method of research which respects the spiritual in nature needs quite other capacities than those which are normally acquired in study or life experience.
First of all, it makes exceptional demands on an exact sense of facts in the quest for truth.
Secondly, capacities must be developed which are akin to artistic observation. What is needed is a sense of relationship, which allows the totality to be seen and experienced in each detail.

Parallel to this outwardly-directed activity goes an inner schooling. An integral part of this is the contemplative experience of the spiritual foundation of one's own thinking-activity. This activity is grounded in the inner life, whereas the outer world can provide only images of what is spiritual.

It is our own being which is able to take hold of an object, separating it consciously from its natural context, that is, making it an object with the help of individual concepts.

One creates a mental picture. Such an introspective path can lead to the assurance that in every mental picture — that is, whenever the self turns actively to the world – the other entity is already inherent. It comes to expression in the name we give to the object, and in the concept, which we have at least begun to comprehend.

In this way we can become conscious of the relationship which prevails between the eternally mobile archetype and our own power to establish connections through thinking.

These new dimensions of consciousness give rise to new responsibilities:

1. When we make our acquaintance with the outer facts in freedom, drawing the criteria by which we judge them entirely from the surroundings under observation (in Goethe's sense – that is to say, not from our fancy or from some abstract order taken to be 'objective').

2. When we place all that we know and do into the whole of a natural and social context.

In former times, through a traditional reverence toward the divine in creation – reflected in even the smallest of actions – beauty and fertility arose in man's environment. Tradition has lost its sustaining power. But in striving to find the totality of relationships for each detail as we experience it, a new kind of reverence may be developed. It may not yet be possible always to anticipate the side-effects of our inventions and modes of action, but certainly we can foresee the directions of their effects.

3. When we establish for our fields of activity clear and surveyable relationships that work back on each other organically, according to the individuation principle in bio-dynamic agriculture.

Efforts in this direction need not be restricted to biologists, ecologists, foresters and farmers. As we have attempted to show in our examples, they can be undertaken and practised by each of us in every life situation.

A training such as we have described is currently being carried out in the one-year Natural Science Seminar of the School for Spiritual Science at the Goetheanum. This is intended as an extension of university studies, and it explores a number of fields. These capacities are also encouraged during the school years from class one to twelve in the Waldorf educational movement, based on the pedagogical methods of Rudolf Steiner. They are applied differently according to age. The fruitful effects of such methods are also evident in agriculture, medicine and other practical fields.

In recent decades there had been widespread reports of damage to forest trees in Europe. The fir tree in particular was prone to illness. Some of the damage was easily traceable to fumes of nearby industries. Since the 1970's, however, the dying has spread in a frightening degree to nearly all kinds of forest trees. The uplands and high mountain areas are most vulnerable, but even lower lying regions have begun to be affected. The situation is similar on other continents (North America, Australia). The fact is undisputed. Since 1980 systematic reports have documented the rapid increase. A new consciousness has been awakened for the essential rôle of the forest in the process of nature and the life of man, and measures are desperately being sought to prevent the complete dying of the forest, as it has already happened, for instance, in the upper Erz Gebirge. The greatest losses have been suffered by the industrial countries, and there seems little doubt that there is a connection with the 'acid rains' caused by industrial pollution.

But the individual symptoms of illness are so various, and the underlying processes which are being discovered so complex, that it is becoming more and more difficult to 'prove' the cause. In spite of this, all experts are basically agreed that there is a connection to environmental pollution.

A 'proof' is not even possible in principle, because science is looking for mechanisms, and has not yet developed adequate approaches to understanding the life of a tree of a landscape, which it generally takes for granted.

Efforts are being made today to reduce the production of pollutants, and the individual can endeavour to help the recovery of the trees through enlivening procedures, such as those developed in bio-dynamic agriculture. But single steps taken by individuals will not in the end reverse the trend towards a loss of vitality in nature. It must be realised that this loss is but the outer expression of a one-sided scientific mode of thought, supported by a naïve faith in progress.

Bockemühl, Jochen, **In Partnership with Nature**. Biodynamic Literature, Wyoming, Rhode Island, 1981

Bockemühl, Jochen, **Manifestations of the Etheric**, (ed.). Anthroposophic Press, New York, 1984

Koepff, Herbert H. (ed.) **Readings in Goethean Science**. Biodynamic Literature, Wyoming, Rhode Island, 1978

Steiner, Rudolf, **Agriculture**. Biodynamic Agricultural Association, 4th Edition, London, 1977

Steiner, Rudolf, **Occult Science, An Outline**. Rudolf Steiner Press, 3rd Impression, London 1979

Steiner, Rudolf, **The Philosophy of Freedom**. Rudolf Steiner Press, 7th Edition, London, 1979

Steiner, Rudolf, **Theory of Knowledge Based on Goethe's World Conception**. New York, 1978

Dr. Jochen Bockemühl, born 1928 in Dresden, studied zoology, botany, chemistry and geology; he took his degree in 1955 with a thesis on the zoology and ecology of ground-dwelling insects. Subsequently, he became co-worker at the Research Laboratory of the Goetheanum, Dornach, Switzerland. His areas of research included: plant development under varying environmental conditions; aspects of leaf metamorphosis; root-ecology and heredity; the question of quality; ecology and behaviour of animals; processes in composting; landscape ecology and landscape architecture; questions of scientific method.

Since 1971, he has been the head of the Natural Science Section of the School for Spiritual Science, Goetheanum (see p.25). He has lectured widely in the English speaking world and run many workshops aimed at developing the ability to observe sensitively the processes of development found in the natural world. Publications include "In Partnership with Nature".